Clayton's Plant Science Series: Plant Anatomy For Babies

Written and Illustrated by Jackie Lemmon

Independently published.

Copyright © 2023 Jackie Lemmon

All rights reserved.

ISBN: 9798374654387

This book was written with love for my nephew, Clayton.

Plant
Anatomy

Plant anatomy looks at all parts of a plant.

Leaf
Shoot
Plant
Roots

Plant

The main Parts of a Plant are the Shoot, leaves, and roots.

Leaf
Shoot
Plant
Roots

Leaf
Shoot
Plant
Roots

Leaf

Leaves collect
Sunlight, which
the plant converts
to food. This
process is called
photosynthesis.
The leaf has a
midrib, many veins,
and a petiole.

Sun
Vein
Vein
Midrib
Petiole
Leaf

Leaf
Shoot
Plant
Roots

Shoot

The shoot houses the vascular system, where the plant transports water up from the ground. Water is transported through the xylem.

Shoot
Xylem
water
Leaf

Leaf
Shoot
Plant
Roots

Roots

Roots are the below ground part of the plant. The roots collect water from the soil, which is transported through the xylem.

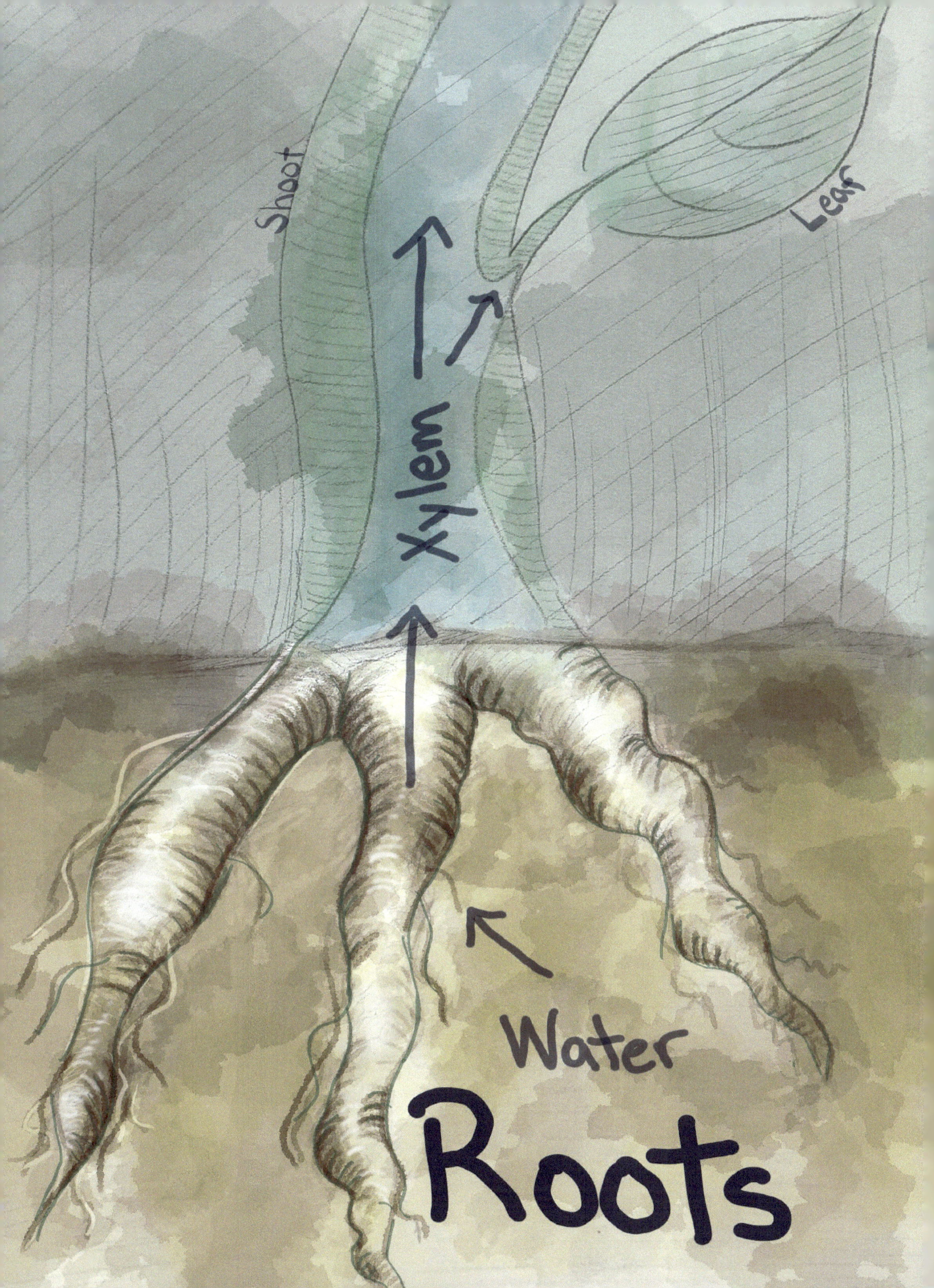

Shoot
Leaf
Xylem
Water
Roots

About the Author

A couple of years ago, Jackie Lemmon became an aunt for her first nephew, Clayton. Hoping to instill wonderment and respect for the natural world into her new nephew, Jackie looked to baby books. Unfortunately, there were no educational baby books on plants. She combines her botany degree as well as her years in horticulture with her art, producing Jackie's debut children's book with this unique educational botany collection.